THE SOLAR FOOD DRYER BOOK

THE SOLAR FOOD DRYER BOOK

STELLA ANDRASSY

EARTH BOOKS
A DIVISION OF MORGAN & MORGAN

SUNHOOD® is a registered trademark

© 1978 by Stella Andrassy
All rights reserved.
No part of this book may be
reproduced or translated in any
form, including microfiling
without written permission from
the author and/or publisher.

EARTH BOOKS
a division of MORGAN & MORGAN, INC.
Publishers
145 Palisade Street
Dobbs Ferry, New York 10522

International Standard Book
Number 0-87100-139-X
Library of Congress Catalog
Card Number 78-58434

· Layout and design by Liliane De Cock

Printed in the United States of America

INTRODUCTION	11
DESCRIPTION AND FUNCTION OF THE SUNHOOD	19
Drawings and instructions for assembling the Sunhood	20
The frame	21
The drip channel	22
The plastic cover	23
Drying trays	24
The importance of the open bottom of the Sunhood	25
The importance of the correct sun angle	27
The best location for the Sunhood	28
Instructions for the use of the Sunhood	29
Case hardening can be prevented	31
Automatic pasteurization	32
How long does it take to dry produce	33
How to package and store the dried produce	35
Utmost cleanliness	36
General preparation of fruits and vegetables	37
THE JOY OF SUN DRIED FRUITS	39
How the Sunhood allows you to obtain more natural sugar in fruits	40
General instructions for choosing the best fruits	41
Pears need special treatment	43
Uses of dried fruits	45
Fruit compotes—fruit salads	45

Apricot whip	46
Apricot nut bread	46
Muesli	47
Making jams with sunshine	48
Sun dried apricot marmalade	50
Cherries	51
Cherry Jam	52
Fruit Leathers	53
Nature's sweets	54
Sun concentrated orange juice	55
Good uses for sun dried orange peels	57
Grapes and raisins	59
Brandied figs for festive occasions	61
SUN DRIED VEGETABLES	**63**
Harvesting the vegetables	64
The need for the blanching of vegetables	65
The preparation of some vegetables before drying	67
Drying	70
The reason why vegetables must be dried harder than fruits	70
The use of dried vegetables	71
Sauteed green beans	72
Coleslaw	72
Baked zucchini and carrots	72
Creamed dried sweet corn	73
Delicious soups from dried vegetables	74
Cream of carrot soup	75
Sun riced potatoes	76
Sun magic with tomatoes	79

Catsup 80
　　Tomato slices—powder—leather 81
　　Instant sun rice 82
　　Beans, lentils and peas 83
　　Lentil or split peas soup mix 84
　　Instant Boston baked beans 85
　　Roasted soybeans 87
HOW TO ENRICH FRUITS AND
VEGETABLES 89
　　Basic enrichment methods 90
　　Orange apples 93
　　Mushrooms 97
　　Dried mushrooms in cream 99
SUN DRIED HERBS 101
　　Make your own herb mixtures 104
　　Herb flavorings 105
　　Herb Teas 106
　　Chamomile for health and beauty 108
STORING THE FRAGRANCE OF
SUMMER FOR THE WINTER 111
USE THE SUNHOOD AS A
MINI-GREENHOUSE 115
FINAL WORDS 116
　　Do not use chemical food additives 117
　　Do not sun dry meat or fish 117
　　Comparison of some food drying processes .. 118
　　Result of some winter tests 119
　　Record keeping chart 121
　　Appendix 124
RECOMMENDED READING 125

INTRODUCTION

More and more people are sun drying their own fruits and vegetables, for there are many advantages in doing so. One can select the very best in farmer's markets or in city grocery stores for preserving. Moreover, buyers can obtain large quantities of fruits and vegetables when prices are at their lowest, yet eat them during times when prices are at their highest.

Many have turned to gardening. It is no longer just another hobby. It has become, for millions, a serious necessity, particularly for families with growing children and for the elderly, living on a fixed income.

But, nature has its own way with timing, and fruits and vegetables have a habit of all ripening at the same time, a fact which all growers know all too well.

Suddenly, the home gardener is surrounded with baskets of red, ripe tomatoes, bushels of apples and masses of other valuable crops. It seems practically impossible to preserve all of this bounty. The freezer is full, canning equipment is expensive; besides, home canning is not an easy art to master. Yet we all know, in these days, no food should be allowed to go to waste.

Many consumers today are also deeply concerned about the additives, chemicals, artificial flavorings and other potentially dangerous ingredients placed in almost all commercial foods. By controlling the processes themselves, the ordinary householder can obtain much purer food. As we know, purer, better tasting foods are more nutritious than those which have been commercially treated, especially those which have undergone high heats, chemical treatments, and suffered from mass production. Such processes inevitably destroy taste, take away vitamins and deplete the food of nutrients.

Of all the ways to preserve food, sun drying is by far the best. It requires the least amount of harmful heat, which can destroy nutrients and flavors, and calls for the least amount of waste. No chemicals, no processed sugars are added. In fact, nothing at all is added. Beautiful, clean sun light, with its germ killing rays is all that is needed. Beneath them, fruits and vegetables dehydrate, they gain more and more concentrated nutrients. Pound for pound correctly dried fruits, for example, contain more proteins, fats, carbohydrates, natural sugars and often more vitamins, minerals and other nutrients than fresh fruit. Interestingly, the drying process itself allows fruits to change carbohydrates into sugars. Anyone who has placed pears or other fruit in a sunlit window to ripen will be familiar with this. All in all, there is every reason why sun drying is the preferred method for preserving fruits and vegetables.

Today there is an energy crisis. Each day the government and news media warns us that we must save on oil, coal, and electricity, and as far as possible replace them with renewable forms of energy such as solar energy. Sun dried fruits and vegetables do not use up any valuable non-replaceable fuels. Since they dry in the sun, all they use is solar heat, which is available as long as the sun shines.

When we think of it, it is truly remarkable how much fuel is used in the preparation of most foods. There is always the transportation of foods from farms to processing plants and from them to the cities. Right here, by the way, the home gardener makes a savings in gasoline bills! There is a further cost of transportation to and from markets. Added to these are fuel costs for processing, refrigeration, packaging, etc. If a person decided to can fruits and vegetables he will inevitably use a great deal of heat in boiling water, etc. This always consumes energy such as gas or electricity. Freezing food initially consumes less fuel, but it is not without its costs as well.

All refrigeration equipment is powered by motors run on gas, electricity or other fuel. Moreover, a freezer must constantly run for months at a time. For these reasons, people who are conscientious about saving fuel, about atmospheric pollution and its growing threat to our environment will do well to consider sun drying for the preservation of their foods. It is the best ecological method.

Sun dried foods, not only have a high nutritional value, but they taste delicious. Those carefully prepared at home are superior to those bought in a store, which often taste of sulfur dioxide or other chemicals used in commercial processes. Most sun dried foods also taste better than canned goods. For example, there is a remarkable difference in the true, genuine taste of dried apricots, compared to those whose lack of taste is hidden under heavy sugary syrups in a can. Solar made catsup, purees, etc. are noticeably better than those in bottles or cans. Delicious soups can be quickly and easily obtained from sun dried vegetables. What could possibly be more nutritious or taste better than home made snack bars made of sun dried figs, apricots, honey, nuts, etc.? In addition, certain herbs, seeds and mushrooms can only be preserved through drying processes, and those which are sun dried once more taste the best. Thus, for flavor, try sun drying.

In spite of the many advantages, many people hesitate to consider sun drying their own fruits and vegetables. With good reason, for what comes to their mind is a mental image of fruits and vegetables lying about on sunlit racks, where they can collect dust and pollutants, not to mention swarms of insects and perhaps ravenous birds or rodents. They also realize that almost any passing cloud brings with it the threat of a ruinous rainfall.

Thanks to the modern Sunhood, all of that is in the past. Worry no longer. There is an answer. Dry your vegetables and fruits in our Sunhood, with controlled sunshine.

The Sunhood, which is a type of solar drier, is placed like a cover over the product which is to be dried, securely protecting it from rain, insects, rodents, dust, etc. It has been developed by me, and its remarkable efficiency has been tested by experts for many years.

The Sunhood is so easy to build that any young or old "do-it-yourselfer," could put one together in no time, provided that the instructions given in this book are followed.

The Sunhood is specifically designed for drying vegetables, fruits, berries, herbs and mushrooms, as well as for concentrating the pulp and juices of various recipes.

The Sunhood, 19" x 25", is not a large device, and it can be utilized by those in cities who do not have much space. In fact, it is ideal for the urbanite, as well as for people who live deep in the country. It is light in weight, yet, it is so sturdy that season after season it can be used. It can remain outdoors overnight and during rainfalls and not suffer any harm.

When the Sunhood is not needed for drying vegetables or fruits, it can be used for drying strawflowers, seed pods, colorful foliages, and fragrant flower petals for potpourris. In the early spring and late autumn it can be turned into a "mini-greenhouse."

One might think, at first, that a Sunhood could only be used in some dry climate, such as the one found in the Southwest. Though it can be used there, it has been used in much less favorable climates. Most of the testing of the Sunhood has taken place in New Jersey, not far from New York City, at a latitude of 41°N, where the rainfall exceeds 40 inches a year, and where even in July the sun only shines about 65% of the possible time it could shine.[1] Few places in the United States have a poorer climate for sun drying foods. Yet, the sunhood has been used up until late November with excellent results. The heat of the sun and its drying powers are far greater than most of us would guess. The Sunhood can be used throughout the United States south of the Canadian border, in parts of Alaska, as well as in Southern Canada.

The Sunhood is an item which will literally pay for itself, over and over again. According to many experts, who have conducted cost estimates on methods of preserving foods, sun drying is the least expensive. For example, Dr. G. A. Shuly, Chemist at the University of Tennessee, has pointed out that home drying of foods cost almost a third less than canning.

Properly dried foods keep well. Many sun dried foods have kept for over a year, while retaining their food value and most of their vitamins.

Anyone interested in ecologically sound practices and who is interested in preparing his own healthy foods will want to make use of the Sunhood.

1. *The statistics come from GOODE'S WORLD ATLAS, Rand McNally & Co., Chicago.*

DESCRIPTION AND FUNCTION OF THE SUNHOOD

The Sunhood is a modern solar drier, which I have invented and which has been extensively field tested by me. Though it is a very simple device, which is easy to make, it has given me excellent results. The techniques and recipes throughout this book are based upon it.

The main parts of the Sunhood are:
1. A wooden frame, 19"x25", with a v-shaped notch.
2. A transparent plastic sheet, attached to the top of the frame.
3. A drip channel, inserted across the middle of the frame at the lowest part of the v-frame.
4. The bottom of the frame is left *open*.

This simple and easily made device is, in fact, a *small solar still,* and under the Sunhood a slow, natural distillation process takes place. Due to the heat of the sun, the water content in the fruits or vegetables begins to distill. The products actually exude water from their tissues. Vapor rises and condenses against the sloping inside of the plastic. Gradually, larger drops form and trickle down into the drip channel, where the water automatically runs out and can be collected in a small jar.

Under the Sunhood the products dry in their own aromatic atmosphere under sterile conditions and pasteurizing temperatures. The constant high humidity, at the beginning of the process, has a very beneficial effect on the texture of the product, as it prevents the outer tissues from hardening.

You can tell when the products are dry and ready for removal from the Sunhood. When no more droplets appear on the plastic, the product is dry.

DRAWINGS AND

INSTRUCTIONS FOR

ASSEMBLING THE SUNHOOD

For the Sunhood frame buy lumber: 6 x ¾ inch, and 8 ft. long. Follow drawings of the design.

After the frame is assembled with screws, it should be painted inside and outside with matte, non-leaded, black paint.

Insert and fasten the drip channel.

Attach the plastic sheet to the top of frame.

Make the inside tray, which does not need to be painted.

THE FRAME

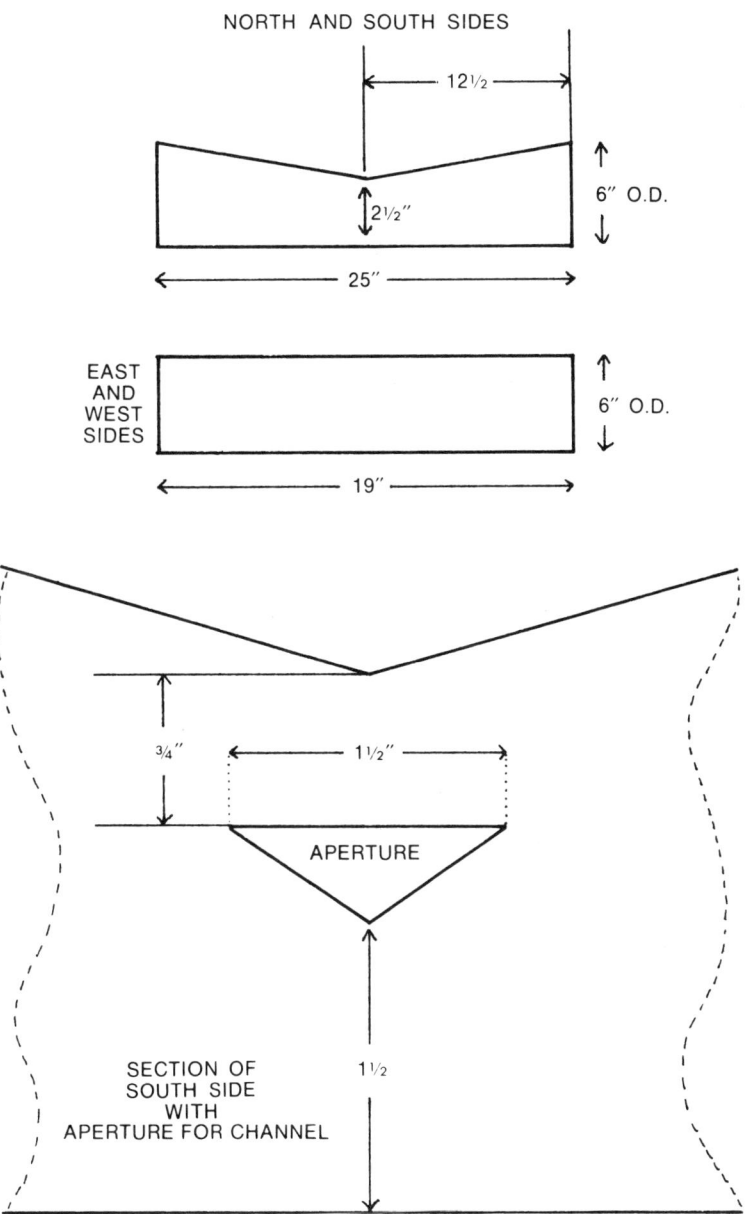

THE DRIP CHANNEL

Use a thin gauge aluminum for the channel. The channel should be V-shaped. About 20 inches long and 1½ inches wide. It protrudes through the aperture on the south side, and is fastened with nails near the top on the inside of the north side.

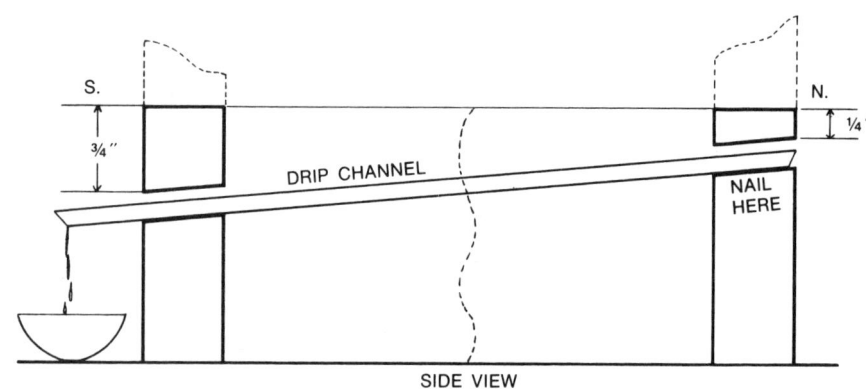

SIDE VIEW

THE PLASTIC COVER

Buy a transparent vinyl plastic sheet, of 4 mil. thickness.

Cut the plastic sheet to size, slightly overlapping the top of the frame.

Fold the overlapping part inwards, in order to strengthen the plastic edge.

Fasten the folded edge of the plastic to the top of the frame: start on the east or west side, press in a few thumb-tacks, and work downwards in the "V".

This operation is easier if you have a helper, for the plastic should be stretched as tightly as possible.

Use the thumb-tacks sparingly to begin with, because you may have to adjust the sheet several times.

When perfectly stretched, staple, or use more thumb-tacks.

As a final touch, attach a tape around the edge, to make the Sunhood more vapor-tight.

During the drying operation water droplets will accumulate on the inside of the plastic. If you tap the plastic lightly on the outside with your fingers, you can induce the water to run down more quickly into the channel.

DRYING TRAYS

The trays should be made in a size that fits easily inside the Sunhood. Trays with sides are better than those without, because the edges keep the food from slipping off when the food is brought from the preparation locality.

Materials: pine wood or unpainted scrap lumber (beware of any poisonous paints such as those containing leads!), acid resistant screen wire or nylon mosquito netting.

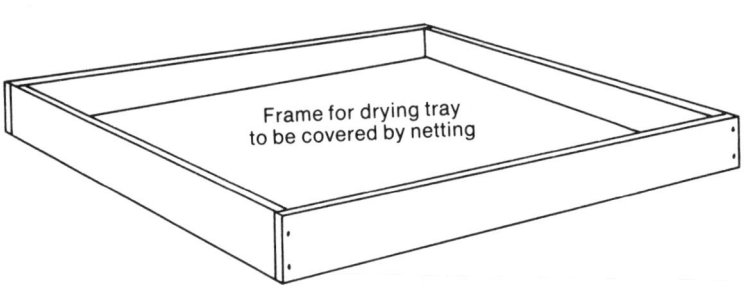

Frame for drying tray to be covered by netting

THE IMPORTANCE OF THE OPEN BOTTOM OF THE SUNHOOD

All products should be turned, shuffled or stirred once a day or when needed. With this model, because of its open bottom, this is an easy operation: With one hand, lift up one side of the Sunhood, and with the other hand, stir or turn the product. Afterward, close the device again. It is that easy.

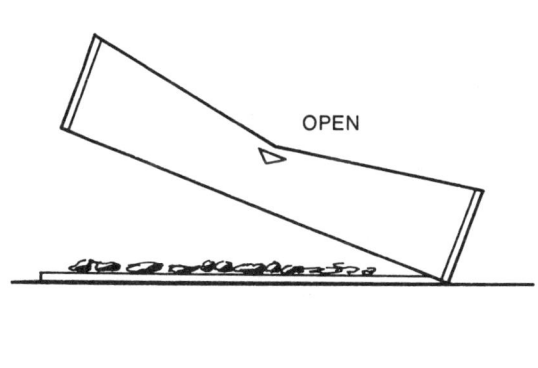

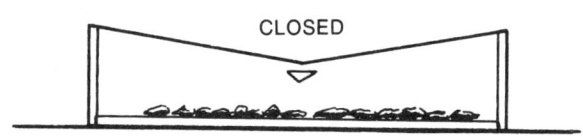

It took me, however, many years to find this simple solution. The first experimental solar driers that I built had either doors or drawers; some even had shelves, which made the operation cumbersome and slow. Moreover, it required a longer time to build such driers.

In comparison, the present Sunhood is so simple in design and so easy to build, that anybody who can handle a saw and a hammer, can build one in a short time.

It is also my hope that the simplicity of the construction and use of the Sunhood will encourage parents to involve their children in its use at an early age.

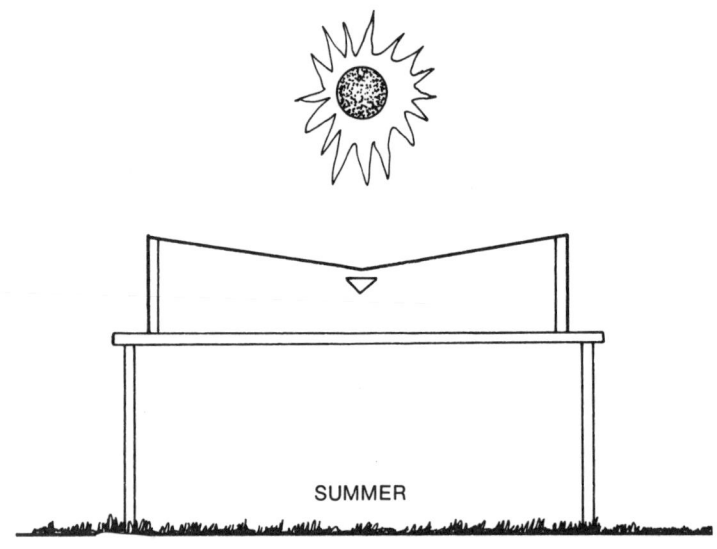

THE IMPORTANCE OF THE CORRECT SUN ANGLE

During the summer months, at noon time, the sun is at its highest point for the year. The Sunhood can then conveniently be kept standing on a table or on another flat surface.

However, in the spring and in the autumn, the sun takes a much lower path. During those months, it is recommended that the Sunhood be placed on a slanting roof or slope of a hillside. If you use a table, prop up the back legs with bricks or logs, etc. so that the Sunhood faces fully toward the sun.

SPRING AND FALL

THE BEST LOCATION FOR THE SUNHOOD

To facilitate the drying operation, choose a sunny spot of your garden, where the Sunhood will be protected from the wind.

For the best results place the Sunhood on the ground or on a table, tilted slightly toward the south. Cover the table or the ground with a clean sheet of plastic.

Before starting, check the sky! Listen to a weather report. If rain is coming do not start your solar drying that day. Remember, the fuel you are using is sunshine and nothing else.

The Sunhood can, however, remain out at night or in the rain. If strong winds are likely to come up, tie it to the table.

INSTRUCTIONS FOR THE USE OF THE SUNHOOD

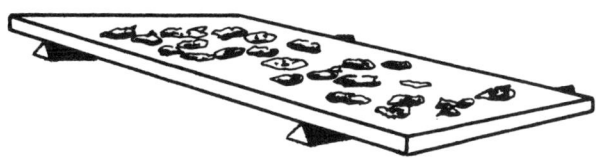

Spread the prepared fruits or vegetables in a *single layer* on a rack or on an absorbent paper, such as kitchen towels.

Due to the heat of the sun, moisture will condense out of the product, small droplets will form on the inside of the plastic, and the liquid will run down and out through the drip-channel.

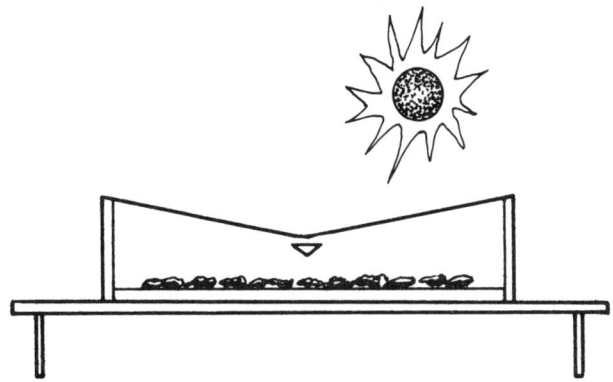

The vegetables and fruits should be turned once a day, or when needed.

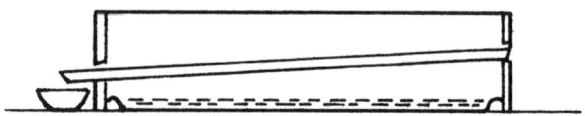

If you are concentrating apple-butter, tomato puree, ketchups, etc., pour them into black shallow pans and place under the Sunhood, where the mush can be concentrated to the desired thickness. Treat fruit juices, marmalades, and jams the same way. STIR OFTEN! The pure water, which exudes from the product, can be collected in a small jar outside the Sunhood.

When no more water droplets appear on the plastic, you can be sure that the product is *dry*.

CASE HARDENING CAN BE PREVENTED THROUGH THE USE OF THE SUNHOOD

One undesirable condition that may be encountered during drying, is the so-called, *case hardening*. This occurs when the outer tissues of a product dry more rapidly than the inner, thus forming something that is like a tough skin, or crust. This hard crust retards the evaporation from the wetter inner tissues, where molds can, and do, develop.

The remedy against this calamity is humidity. To provide this, a certain amount of steam is introduced into Machine Driers during the dehydration process.

Now, fruits and vegetables, which are dried under the Sunhood, *do not get case hardening* because during the initial stage of the drying process they are automatically surrounded with a protective vapor barrier.

At night, although there is an obligatory interruption of the drying process, it is not completely halted, and the emission of moisture continues, though at a slower pace than during daytime. This retarded emission of moisture during the night, has a certain favorable effect because it permits a moisture exchange between the outer, drier tissues, and the wetter, inner tissues. Thus, an equalization sets in, and the entire drying process remains constant, and mold formation due to case hardening is prevented.

AUTOMATIC PASTEURIZATION

During the daytime, the temperature under the Sunhood varies from 130 to 150 degrees Fahrenheit, which is the acceptable temperature for pasteurization. Furthermore, the sun's rays have a deadly effect on most bacteria. When the product is ready for storage, it is therefore a safe and hygienic product; as long as it is under the protective cover of the hood, no new contamination is probable. It is only during packaging and storage that new contamination can occur. Therefore, as an added precaution against spoilage, it is recommended that the dried product be pasteurized again when stored in its container. Place the closed containers under the Sunhood and leave for one half hour on a bright sunny day.

HOW LONG DOES IT TAKE TO DRY PRODUCE?

"How long will it take to dry my fruits or vegetables?" is a question I often hear. My answer always is: "That depends entirely on the weather. Sometimes a few hours is sufficient, but other times, a few days are required."

It is understandable that a modern person, endowed with a "push-button" mentality, will find it difficult to accept the fact that one can never know exactly when a product will be ready. Yet, those who want to work with sunshine, must learn to adapt themselves to the *sun's own timing*. It is often unpredictable but, in the long run, it is very *reliable*.

It is, therefore, impossible to prescribe the exact drying time for any product, since so much depends upon the weather, the geographical and physical location of the Sunhood, and the product itself. You must use your own judgment and observe the water droplet development on the inside of the plastic, as well as test the product for the desired dryness.

Following are a few examples of how the drying time can vary in New Jersey—forty miles south of New York City:

Grapes—six days in July, but fourteen days in November.

Apple slices—one day in June, but three days in November.

Spinach, parsely, dill, celery tops, etc. and all

herbs—require only a few hours for drying in good sunshine. When drying herbs, in order to better retain their color, it is advisable to place the Sunhood in open shade.

Open shade can be defined as an area underneath an overhang—near the edge of a porch—underneath a parasol—any place where the Sunhood is not exposed to the direct rays of the sun, but exposed in such a manner that it catches the drying warmth of the sun.

HOW TO PACKAGE AND STORE THE DRIED PRODUCT

There are a few ironclad rules which must be observed in order to keep dried fruits and vegetables in good condition during storage. They are: CLEAN, DRY, DARK, AND COOL.

Before you package the product make sure that it is thoroughly dehydrated. The importance of this cannot be overstressed. If any moisture remains bacteria and mold WILL develop and grow.

The dried product should be stored as soon as possible after dehydration is completed, however, allow sufficient time for the fruits and vegetables to cool off. If they are enclosed while warm, moisture can develop.

Storage containers: Metal containers (such as coffee or shortening cans) can be used after thorough scrubbing and rinsing. Be sure that no odors remain. Plastic containers and glass jars are also suitable. For all, tight fitting lids should be used. Glass jars are particularly recommended in humid climates.

Heavy plastic bags, especially the sealing freezer kind, are good. When using the non-sealing type remove as much air as possible before closing.

Many small containers are better than a few large ones.

Plastic and glass does expose the contents to light, so either store them in brown bags or dark places.

Despite all your precautions, things can go wrong. It is a good idea, especially in long term or large quantity storage to occasionally check the containers. Remove all pieces with even the faintest signs of mold. It is advised that any contaminated batch be redried. If it is a sunny day, spread the produce on the tray under the Sunhood and dry for about one hour, then restore. If you discover mold in the middle of winter and no sun is in sight, dry in a 150-175°F oven for 10-15 minutes.

UTMOST CLEANLINESS

Wash all vegetables and fruits very well before they are cut because drying *increases the percentage of spray residues* and other impurities in proportion to the drying rate. Remember, DDT and other related compounds are still used.

Cleanliness is the watchword in all phases of the drying production.

Here are some things to keep in mind.
1. Wash your hands often.
2. Use only clean and sharp knives.
3. Make complete cuts. Do not break or tear fruits.
4. Keep all the pits separate. Do not let them fall onto uncut fruit.
5. Throw away . . .
   a. fruit showing rot or mold
   b. fruit showing insect infestation
   c. fruit with imbedded dirt.
6. *Clean the trays after each use.*

GENERAL PREPARATION OF FRUIT AND VEGETABLES

The process of preparing fruit and vegetables for sun-drying differs somewhat for each type. There are, however, certain general rules, which apply to *all of them*. They are:
1. *Select the best.*
2. *Keep the utmost cleanliness.*

No fruit or vegetable or part of one should be dried which would not be considered edible in fresh form. Before starting the preparation for drying, ask yourself: "Would I eat this fruit or vegetable as is?" When a dried fruit or vegetable is cooked or eaten as is, it cannot become better, than the fresh one you started with. Worm-infested, or rotten fruit, old, dried or mushy vegetables, etc. must therefore be discarded at the beginning of the preparation.

Remember this rule: the best dry best!

THE JOY OF SUN DRIED FRUITS

Of all plant products fruits are among the sweetest, most nutritious, and flavorful. This section will tell you all that you need to know about the selection and drying of fruits. It will also describe new methods for combining vitamins and minerals into fruits and even making new products, such as the orange-apple. Many will want to make fruit bars and other natural sweets for children and for themselves, especially as they are so great for munching.

The detailed, clear instructions will make it all easy for you. It is best if you read all of the instructions so that you can obtain valuable hints and ideas as you go along.

HOW THE SUNHOOD ALLOWS
YOU TO OBTAIN MORE
NATURAL SUGAR IN FRUITS

Now there is yet another significant advantage of drying fruits under controlled solar conditions, instead of using the outmoded practice of open-air drying. Since the owner of a Sunhood is no longer so dependent on the weather, he can dare to *delay the harvest* until the fruits on vines and trees have reached their peak of ripeness.

This prolonged ripening period increases the sugar content of the fruits considerably, which automatically upgrades the product. Thus, later harvests and longer ripening periods result in a heavier and sweeter product, which in turn fetches higher prices in the market.

Statistics show that by delaying the picking of grapes from August 15 to September 16, there was an increase in weight of 43%. The sugar content of the grapes during the same period increased from 21% to 28.8%.

GENERAL INSTRUCTIONS FOR CHOOSING THE BEST FRUITS

It is important to obtain the very best fruits at the right time. The following are general instructions for harvesting various fruits. If you do not own or have access to a garden, try to obtain the freshest, ripest that you can get. Your best bet is to go to farmer's markets. If you live in a large city, shop early when the freshest produce is arriving in the shops.

Harvest all fruits early in the morning or late in the afternoon. Choose only the best and flawless fruits.

After preparing the fruit, spread in a single layer on the tray. When dried, fruit will feel leathery and pliable.

Apples:
Should be picked when ripe, but not mushy. They should not be bruised or deformed.

Apricots:
Should be harvested just before the fruits drop from the tree. *Nectarines* and *peaches* are harvested the same way.

Figs:
Should be left on the tree until they are ripe enough to fall to the ground. If unripe figs are dried, the product will be woody and tasteless.

Pears:
One of the best kinds of fruit for drying is the Bartlett Pear. Harvest these pears when they are still firm, yet easily picked. It is a good custom to let pears *after-ripen,* in a covered crate, or box, for about a week before preparing them for drying.

Prunes:
Should be allowed to ripen on the tree until they fall to the ground.

Blueberries:
Should be picked early in the morning, before they get heated by the sun. When picking berries for drying, use a wide, shallow basket or a plastic lined box, so as not to crush the berries. Always try to dry berries the same day they are picked.

Strawberries and raspberries:
Are not suitable for drying, but can be used in sun dried jams.

Cranberries:
Dry well, cut in halves, before drying.

PEARS NEED SPECIAL TREATMENT

Pears are a flavorable dried fruit, which most people love. The preparation of them for drying involves processes not used for other fruits. They are:

Picking:
    Pick or buy pears when they have reached full size, but still are hard and green.

Ripening:
    Keep the pears for after-ripening at room temperature—for about a week. Sort out periodically. *Pears are ready for drying when they are eating-ripe.*

Spray residues:
    Before the pear is cut, it is necessary to remove all spray residue, if such has been used, through thorough washing. The presence of the calyx makes the pears more difficult to clean than other fruits. The calyx or the blossom is best removed with a calyx remover, and the stems are pulled out.

Cutting and trimming:
    The pears are cut in half lengthwise and trimmed to remove all damaged parts. It is not customary to peel pears, nor to remove the core, unless it is damaged.

Washing and traying:
    A second washing is done before the traying, in order to remove bits of trimming and insect ex-

creta. Place the fruits cut side up on the trays and place under the Sunhood.

Test for dryness:

When ready, the dried pears will be flat and flexible, not mushy, with little browning or curling of the cut edges.

USES OF DRIED FRUITS

Fruit Compotes:

Dried apricots, peaches, pears, etc., can be eaten as is, or soaked in water and kept in the refrigerator overnight. When dried fruits are to be used as compote, then some lemon or orange juice added to the water helps the taste. Sugar should be used sparingly and added only *after* the fruits are cooked.

Fruit Salads:

Minced dried figs, crushed nuts, and dried shredded coconuts are interesting additions to any fruit salad.

Here is a recipe of a Syrian fruit salad, called *chosaffe*, which is made of dried fruits:

4 oz. prunes
4 oz. apricots
4 oz. figs, or any other dried fruit available
2 oz. raisins
2-4 oz. nuts
2 oz. almonds
3-4 tablespoons sugar or fruit syrup
1 tablespoon rosewater (optional)

Leave the dried fruits to soak in cold water in the refrigerator or cold place for 24 hours. Soak the nuts in another bowl.

Blanch the almonds, remove the skins and add the nuts to the fruits. Sweeten, if needed, with sugar and rosewater. Serve very cold.

Apricot Whip:
>1 cup dried apricots
>½ cup sugar
>2 tablespoons lemon juice
>¼ teaspoon grated lemon rind
>¼ teaspoon salt
>3-4 egg whites

Soak the dried apricots in water for several hours, or overnight. Simmer under cover until very soft. Cool and press through a strainer or food mill. Add half the sugar and cook until thick. Cool slightly and add the lemon juice, grated rind and salt. Fold in the egg whites beaten stiff with the other half of the sugar, which is added spoon-by-spoon during whipping. Pour in a buttered baking dish and bake about 45 minutes at 350 degrees.

Variations:

Other dried fruits (prunes, peaches) can be used instead of the apricots.

Apricot Nut Bread:
>½ cup diced dried apricots
>1 egg
>2 tablespoons melted butter
>2 cups flour
>3 teaspoons baking powder
>¼ teaspoon baking soda
>½ teaspoon salt
>½ cup orange juice
>½ cup water
>1 cup walnuts, broken in pieces

Soak the apricots for several hours. Drain and grind. Beat the egg with the sugar until foamy. Add the melted butter and mix well. Sift the flour with the baking powder, salt, and baking soda, and add alternately with the orange juice and the water to the egg and sugar mixture. Finally, fold in the apricots and the chopped walnuts, mixing carefully.

     Bake in a 300°F. oven for about 1½ hours until dry when tested.

Muesli:
Muesli is the famous health food which originated in Switzerland. It is particularly recommended for growing children, athletes, and people who have to perform hard physical work. It is ideal for campers and for having in reserve in the home for young people who desire a night snack.

     For one serving, take 5 tablespoons of oatmeal and soak in milk or fruit juice for several hours, or overnight. Sweeten with honey to taste.

     Dried fruits, such as raisins, apple nuggets, diced apricots, etc., should be added after they are soaked. Grated nuts, or grated almonds also belong in a Muesli. Fresh fruits can, of course, also be added, if available. It is a very individual health food and each person can mix in it as much or as little of each ingredient as desired.

     It is advisable to keep each ingredient in a separate jar or plastic bag and not premix raisins, apples, nuts, etc., because they may not keep well in storage when mixed. Powdered milk can be substituted for fresh milk.

MAKING JAMS WITH SUNSHINE

The finest tasting jams in the world are those made through sun drying processes. They beat all others by a wide margin.

Strawberry, raspberry, peach and other jams
Many fruits, like *strawberries, raspberries, peaches,* etc., contain too much water to make a good jam without adding a thickener or using prolonged cooking, which destroys much of the fragile flavor of the fruits.

For generations superior cooks have known the art of making strawberry jam under glass, using only sunshine as fuel.

What we now suggest is a new version of this ancient method:

In order to remove some of the unwanted water, strawberries, raspberries, etc., are placed in a single layer in a pan under the Sunhood *without* any sugar. Evaporation will set in and the fruits should be allowed to shrink—but not to dry. At this stage, the pre-measured sugar is poured over the fruit and will melt in the remaining juice. The amount of sugar will depend on the tartness of the fruit and your taste. A good start is ¾ to 1 cup of sugar to each cup of fruit pulp. The jam making then continues under the Sunhood, with some stirring, until the desired thickness is obtained.

This method of driving out some of the water from the fruits before the sugar is added, can be used not only for all kinds of jams, but also for fruit compotes.

Apricot Marmalade:
This is an easy way of using dried apricots, which does not require straining of the pulp.

Soak the apricots in water overnight in a china or glass bowl. Cover the fruit with 3 to 4 inches of water.

Next day boil the fruit in the same water for about one hour. (Use an enameled or heat proof glass pot.) Boil until the apricots fall apart. Stir the fruits and crush them with a silver or wooden spoon.

When the pulp appears sufficiently soft, separate it from the juice and measure it. To 1 cup of pulp add 1 cup sugar, add the juice and the grated rind of one lemon.

Mix the sugar into the hot pulp together with the lemon juice and the remaining juice. Return to the fire.

Boil this mixture gently for about 20 to 30 minutes, stirring carefully to prevent burning. When ready the marmalade should have acquired a deep reddish brown color.

Pour the apricot marmalade immediately into sterilized jars. When cooled, cover with paraffin.

CHERRIES

In my little garden there grows a cherry tree, which in early spring, when it is covered with blossoms, is a sheer delight. This lovely tree also carries a lot of cherries, which alas, are of an indescribably sour variety.

Nevertheless, since one of the principles of my life is: "Never let any food go to waste," I decided to make these cherries palatable.

A basket full of cherries was crushed in a big enameled kettle. No water was added, but there was plenty of juice for the cherries to boil in. The boiling was continued until the fruits fell apart. The pulp with the stones was then poured into a cheesecloth and the filtered juice collected in a bowl.

One quart of sugar was added to each cup of juice. This rather liquid mixture was then poured into the black enameled pans (about ½ inch deep in each pan) and placed under the Sunhood, to allow some of the water to evaporate.

At the end of the first day, the juice had turned into a thick syrup (about the consistency of maple syrup), but by noon, the second day, it had been converted into a dense delicious jelly, which was promptly packed in glasses and while still warm, covered with a layer of paraffin.

Cherry Jam

    4 cups of pitted cherries

    2 cups of sugar

Let the mixture stand overnight to release the juice. After a quick boil, the rather liquid cherry jam is poured (in a layer of about 1 inch), in a black shallow pan and placed under the Sunhood, where it is kept for two days, while the jam is stirred a few times. Towards the afternoon of the second day, you will have a thick cherry jam.

FRUIT LEATHERS

Many fruits can be made into fruit leather. Some that work best are apples, berries, peaches and apricots.
      The selected fruit should be well ripened. Wash thoroughly, remove the pits or cores, and cut into small chunks. Peeling is optional. Boil in the least amount of water possible until mushy. Sugar or honey can be added before boiling but this is optional and depends on your taste and the tartness of the fruit.

After boiling let the fruit stand until cooled and then puree it by pushing it through a fine meshed sieve or use a food mill.

Line the drying pan with strips of plastic freezing wrap and oil the plastic. (The strips should overlap the sides of the pan so it is easier to remove.)

Pour the puree in the pan and smooth out to a thin layer, about ⅛" thick. Put to dry under the Sunhood. When dried, the leather will be pliable and the surface will feel slick. Peel the leather from the plastic and cut into strips, squares or whatever. Store into dry, well closing containers or wrap in plastic and store in the refrigerator. For longer keeping, fruit leathers can be stored in the deep freezer.

NATURE'S SWEETS

Fruit bars:

Take equal amounts of finely chopped dried fruits. This can be a combination of figs, prunes, apricots, peaches, raisins, or whatever you have on hand. Mix well and bind with honey. Shape in bars about 1" x 1" x 3" and wrap in cellophane for storage.

These bars make a great and healthy snack for campers, hikers, fishermen, not to speak of children!

SUN CONCENTRATED ORANGE JUICE

It was as late as the middle of November, when I made my first experiment of concentrating orange juice in *their own peels* in order to enhance the flavor. The oranges were cut in half and placed with the cut side up under the Sunhood. (I kept an equal amount of oranges for control.)

At this location, in the middle of the East Coast, the sun is not strong in November and the nights are chilly. But in spite of this, on the second day when I squeezed one of the solar exposed oranges, I found that:

a. It released the juice much easier than the control.

b. It was much sweeter, and had obtained a richer taste and flavor.

c. It yielded in proportion more juice, than the oranges which had not been solar treated.

The explanation to this transformation may lie in the fact that the orange juice which is being concentrated in its own aromatic environment absorbs the flavoring elements from the peels and the volatile oils which usually escape from them. The increased sweetness is due to the effect that during drying, some of the carbohydrates (starches) change into sugar.

Time:
The time required for this process depends naturally on the fruit itself, the location and time of year. If you want to concentrate orange juice in Florida in the middle of July, you may need only half a day for obtaining desired result. Best try it out first on a few oranges.

It is a known kitchen trick that if you drop an orange in boiling water, it becomes easier to peel or squeeze. The same result can be obtained if the oranges are warmed up under the Sunhood.

Freshly pressed orange juice can be placed in shallow utensils and reduced to the desired concentration for freezing. Grated orange peel can be added if desired.

GOOD USES FOR SUN DRIED ORANGE PEELS

Nobody can count the millions of oranges that are squeezed for juice in the American household each morning. Nor is it possible to guess how many orange peels are being thrown into the garbage cans. What a waste of good material, which could be used in so many ways!

To dry the peels: Wash them first and remove some of the white pulpy matter. Cut the peels in chunks—slice them or chop them. Orange peels dry very quickly. After being dried, they should be kept in tight closing jars or plastic bags.

Dried orange peels add a delectable flavor to many foods, as an example:

A few slices of dried orange peels, placed in the cavity of a chicken or small turkey, before roasting give a piquant taste to the gravy.

Presoaked chopped orange peels can be used in orange bread and baked goods like Zweiback or crumpets.

Or here's another use: If you are a tea drinker and can't afford the more expensive brands, this is how you can upgrade an otherwise flat tasting tea: To each tablespoon of tea, add one pinch of dried orange peel. If, in addition you drop a clove into the pot, and steep for 7 minutes, then the result will be a fragrant cup of tea, comparable to the very best brand one can buy.

GRAPES AND RAISINS

Among the many varied assortments of material that I have dehydrated, from seaweed to multitudes of fruits and vegetables, no result ever gave me such a satisfaction as when I succeeded to produce a few pounds of raisins, as far north as this location, a short trip south of New York City. It was thought that sun dried raisins could only be produced in warm and dry regions. I was, therefore, somewhat skeptical when I made the first test, to produce raisins from "Thomsons Seedless Grapes" in the late autumn.

Before placing the grapes on the rack, I cut them off in clusters from the main stem. I then washed them thoroughly to clean them of any spray residue and dust. The clusters were than well spaced, from each other, in order to prevent any mold from developing.

It was one of the last days of October, sunny and beautiful, when the test began. But a few days later the weather changed. We got heavy rains, and the temperature dropped close to freezing at nights. But in spite of these adverse weather conditions, the drying process continued, and to my great joy, on November 11, I could harvest the sweetest, driest raisins ever made in this part of the world!

Since that time, I have dried raisins many times, but luckily under more favorable circumstances. This summer, in July, for example, the raisins were ready in five days, with only one rainy day between.

It is important to turn the grape clusters every day, and to remove any fruit that has gone bad. It is, however, remarkable that during all the drying tests that I have conducted, I never had a single rotten grape.

An important factor that probably contributed to the successful drying of the grapes, during that cold and rainy period in November, was the fact that I placed an insulation under the Sunhood. This insulation was made of several layers of newspapers, wrapped in plastic.

Raisins can be dried in Northern Latitudes

The favorable results which were obtained through drying grapes under the Sunhood at this latitude (41°) indicates that *it is possible to push the production of dehydrated fruits further north* into areas which hitherto have been considered unsuitable for open-air drying, due to autumn rains and chilly nights.

Brandied Figs for Festive Occasions:

Here is a recipe for delicious brandied figs:
   25 fine dried figs
   Water to cover
   Juice of one lemon
   4 whole cloves
   1 inch long cinnamon stick
   Red wine or brandy

Take the best dried figs you can find, and simmer them very slowly in the water together with lemon juiuce and spices until the figs swell and become plump. Stir carefully.

When desired plumpness is reached, drain the figs, and place them with the spices in a sterilized jar.

Continue to simmer the remaining liquid until it barely covers the bottom of the cooking pot. While still warm, pour it over the figs, together with the spices.

Fill the remaining space in the jar with slightly heated (not boiled) red wine or brandy, until the figs are completely covered. Close the jar and cool. The longer the brandy figs are stored, the more they improve. The fig wine constitutes a good tonic, and the digestive properties of figs are well known.

Place one brandy fig on top of each individual serving of vanilla ice cream. Pour a little of the brandy over the fig, or serve the brandy in a small glass separately.

A jar of brandy figs can make an elegant, as well as an unusual, gift.

SUN DRIED VEGETABLES

Vegetables are an important, necessary part of your diet. The Sunhood can provide us with a good supply of dried vegetables to use during the long, cold winter months.

In the following section you will find all that you need to know about the simple process of drying vegetables, herbs, mushrooms, etc., and also recipes on how to use them.

HARVESTING THE VEGETABLES

The harvesting and buying of vegetables is very much the same as with fruit. Be sure to harvest them in the morning or late afternoon when they are cool. Always choose the plumpest and the ripest, however, remember that soft spots and mushiness is a definite sign that a vegetable has passed its peak ripeness. They should feel slightly firm to the touch. With a little practice you can always get the best, for only those will give you that great taste treat late in winter.

THE NEED FOR THE BLANCHING OF VEGETABLES

The Importance of Blanching
The purpose of blanching is the inactivation by heat, of enzymes, which are responsible for undesirable changes in color, odor, texture, as well as loss of vitamins.

Blanched vegetables will reconstitute easier and require less time in cooking. Unblanched, dried vegetables tend to be tough and can possess off-flavors and odd odors.

Therefore while blanching is a chore it is well worth the extra effort.

There are two methods of blanching; boiling and steaming. Boiling is preferred for leafy vegetables, the heat penetrates quicker, while watery vegetables, such as beans, squashes, and corn benefit more from steaming.

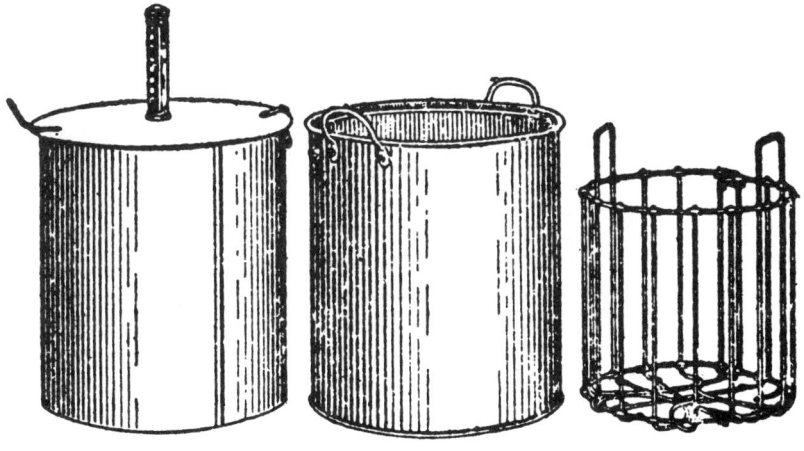

Boiling:

Immerse vegetables in a large amount of rapidly boiling water. A colander or wire basket is very useful for this purpose. The vegetables should be packed, loosely, in a layer not more than 2 inches thick. Vegetables are done when they become slightly wilted, or to test remove a piece from the center of the basket, it should feel tender but not soft. When done, vegetable should be dunked in ice water to stop any further cooking. Chill well, drain and dry on paper towels.

Steaming:

Suspend the vegetables, layered in a wire basket, ABOVE rapidly boiling water. Cover the kettle. Start timing when steam escapes from underneath the lid. When done, proceed as with boiling.

Always use unsalted water for blanching.

THE PREPARATION OF SOME VEGETABLES BEFORE DRYING

Before blanching, most vegetables should be cubed or sliced in thin strips.

Beets:
   Only tender beets should be used. After a good but gentle scrubbing, they should be steamed with the roots and an inch of the top still attached until completely done, 30 to 35 minutes. When cool, remove the skin and then slice, cube, or grate.

Cabbage:
   Either white or red cabbage can be used. Remove the outer leaves, core and cut into shreds about $\frac{1}{8}$ inch wide. Steam two to three minutes, or until the cabbage wilts, but does not collapse.

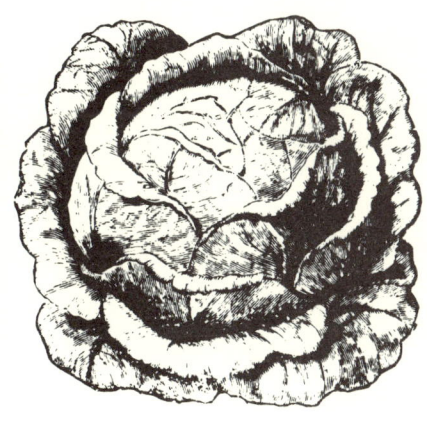 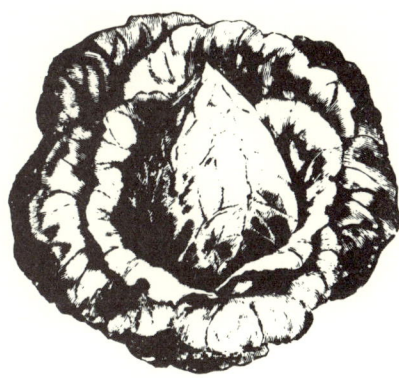

Carrots:
    Wash and trim the carrots, cut into ¼ inch cubes or shred. Steam or boil for about 5 minutes.

Celery:
    Wash well. If to be used for soup, powder or soup mixtures, the entire stalk and leaves are shredded. Steam for one to two minutes. The leaves can also be dried separately without blanching.

Corn:
    Corn is best harvested when just ready for eating. The husked and trimmed ears are steamed until the milk "sets," that is, until no liquid exudes when the kernels are cut with a knife. Ten to twenty minutes is recommended. After cooling remove the kernels from the ears.

Green Beans:
   The ends must be cut off and the strings removed. Slice them length-wise or cut them, on the diagonal, in small pieces. Boil or steam for three to four minutes.

Onions:
   The outer layers are removed and the onions sliced in ¼ inch thick rings. They do not require blanching.

Garlic:
   Remove the skins and cut the cloves very thin. They do not require blanching.

Peas:
   The peas should be young, tender, and sweet. They should not be allowed to stand after shelling because they rapidly lose their sugar and flavor. Steam for five to ten minutes.

Peppers, pimentos and chilis:
   Wash and slice in strips. They do not need blanching.

Squash and zucchini:
   Trim but do not peel. Cut into ¼ inch slices. Steam six to eight minutes.

   For *tomatoes, mushrooms, herbs* and *potatoes* see separate sections.

DRYING

After vegetables have been blanched and cut, spread in a thin layer on the screen, without crowding, to allow the air to circulate. Most vegetables will be brittle when completely dried. Corn and peas will be shriveled and rock hard. Sliced carrots and mushrooms will be leathery and tough.

THE REASONS WHY VEGETABLES MUST BE DRIED HARDER THAN FRUITS

Vegetables must be dried to a much lower moisture content than fruits to keep satisfactorily. This is due to the fact that the starch in the vegetables does not dissolve in the vegetable juice; whereas the sugar of fruits does dissolve. Consequently, on drying, the juice in fruit becomes highly concentrated in sugar content, which prevents the growth of yeast molds and bacteria. The juice of vegetables is thin and low in sugar, hence must be concentrated to a very small volume in order that the resulting syrup will have sufficient preservative power to prevent spoilage by microorganisms.

THE USE OF DRIED VEGETABLES

Dried vegetables can be used in any regular recipe, but first need to be reconstituted. All but leafy vegetables are covered with cold water, with just enough to cover the vegetables. Let stand until they have regained close to their original size and appearance. Add water if necessary, do not drain, but cook in their own liquid. Proceed with regular recipe. Leafy vegetables are covered with boiling water, let stand for five minutes and then use as the recipe requires. Soup mixes and powdered vegetables can be cooked without presoaking.

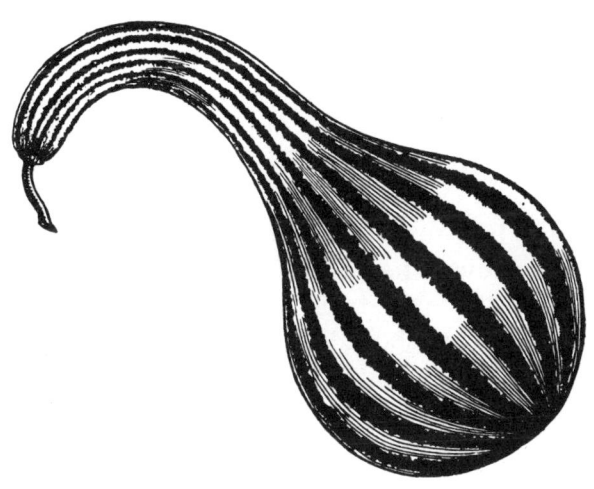

Sauteed Green Beans:

Soak dried beans in water to cover for a few hours. Melt several tablespoons of butter in a frying pan, add chopped onions and green peppers (either fresh or reconstituted dry) and saute until limp. Add the beans and cook for 3 to 4 minutes. Stir in a few tablespoons of sour cream, stir well and serve hot.

The following make nice additions to any green beans dish: sliced mushrooms, slivered almonds, chopped water chestnuts or chopped Jerusalem artichokes.

Coleslaw:

Pour boiling water over dried shredded cabbage to cover. Let stand for a few minutes, then cover and chill for 1 to 2 hours. Add shredded carrots (either fresh or reconstituted dry) and a handful of raisins. Mix with mayonnaise and a few dashes of vinegar.

Baked Zucchini and Carrots:

Reconstitute equal amounts of shredded zucchini and carrots. Add one grated raw potato and chopped scallions to taste. Mix well and put the mixture in a buttered casserole. Beat 1 or 2 eggs and pour them over the vegetables, cover with cornflake crumbs and a few pats of butter. Bake in a preheated oven at 350 for about 45 minutes.

Dried zucchini or carrot slices make great substitutes for potato chips.

Creamed Dried Sweet Corn:

At the first Thanksgiving dinner that I was invited to in America, I was served among all the other trimmings, *creamed dried sweet corn.*

It was in South Bend, Indiana, and my gracious hostess, Mrs. Mabel Ziegler, pointed out to me the background of this dish, which I had never tasted in Europe.

She told me how the early settlers learned from the Indians how to prepare it through cooking the husked ears, scraping off the kernels, and drying them in direct sunshine.

Following is Mrs. Ziegler's recipe.

Soak one cup of dried corn in two cups of water for several hours. Don't drain, but let it simmer in the same water, to which is added one tablespoon sugar, until soft (about one hour). Add more water if needed.

Add salt and pepper to taste, some butter and a few tablespoons of sweet cream and serve.

DELICIOUS SOUPS
FROM DRIED
VEGETABLES

Everybody seems to appreciate dehydrated vegetables best when they are served as soup.

For this purpose the vegetables should be dried in flakes or ground and powdered when bone dry.

You can make your own "ready soup mixes" through drying each vegetable separately and mixing them together only after each is perfectly dried. Such mixtures are very handy for cooking with soup bones or chicken necks, but I found it better to keep each vegetable in its own container and make the soup mixture only when I wanted to cook it. As examples: potatoes with onions or celery, or tomatoes with onion and green peppers, etc.

Cream of Carrot Soup:
> 4 cups of milk
> 6 teaspoons powdered carrot
> 2 tablespoons butter
> ½ -1 teaspoon sugar
> salt, pepper, and other seasoning to taste

Heat the milk, but do not boil, with the powdered carrots and spices in a double boiler and slowly simmer for five to ten minutes. Take off the fire and stir in the butter in small pieces until it melts. Sprinkle with parsley and serve with diced, fried bread.

The same measurements can be used for almost any other soup made of dried vegetables. For variety, exchange the milk for meat of vegetable stock.

SUN RICED POTATOES

In order to make riced potatoes with the aid of the sun, one must take the following steps:

Steam unpeeled potatoes until soft, but not mealy. Peel, and while still warm, press the cooked potatoes through a ricer in a *thin layer* directly on a tray, covered by wax paper. Spread the material with a fork so that no sticky lumps can form. Placed under the Sunhood, on a day when the weather is ideal, the riced potatoes become bone dry within three hours, acquiring a faint golden color. Sun riced potatoes can be used according to any mashed potato recipe.

Instant mashed potatoes are a blessing for any cook in a hurry. But, people who object to chemicals in their food, should read the labels on the boxes of the instant mashed potatoes which are on the market.

Here is an example from such a label: Contains: potato flakes, vegetable Mono- and Diglycerides, Vitamin C, Sodium Sulfites (preservatives), Calcium, Stearoyl-2Lactylate, Sodium Acid Pyrophosphate, Citric Acid, Niacin BHA and BHT (preservatives), Vitamin B_6 Hydrochloride, Vitamin B_1 Mononitrates, Natural and Artificial Flavoring.

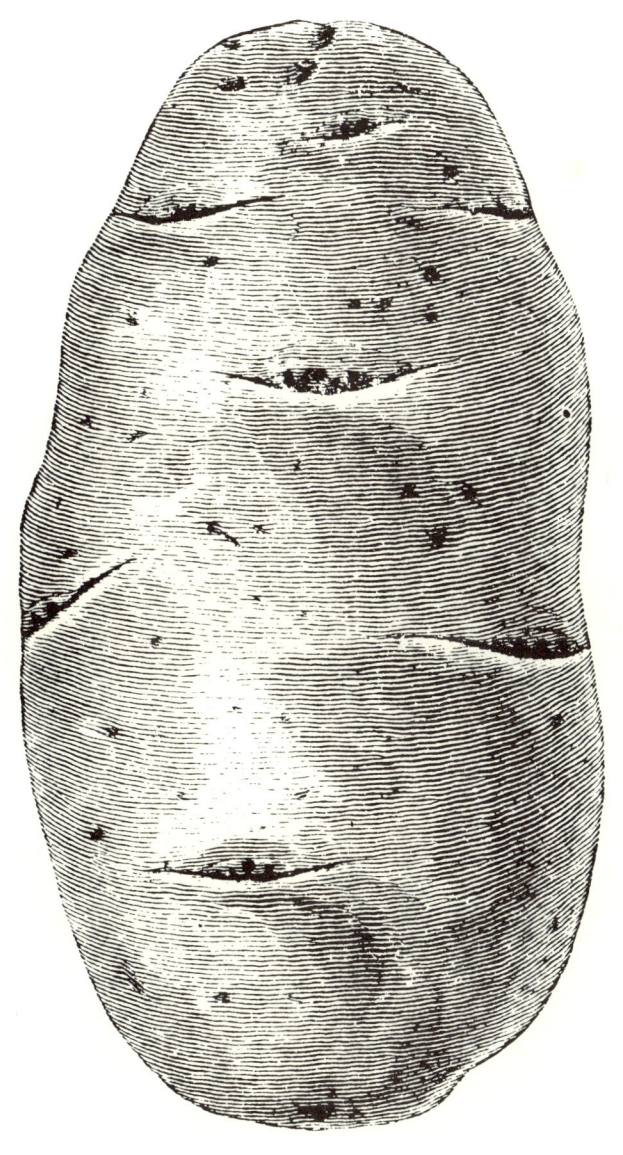

SUN MAGIC
WITH
TOMATOES

The summer has reached its peak and the kitchen-garden is at its very best. The red ripe tomatoes are hanging in heavy clusters on the vine. It is a joy for the eye of the gardener, but also a reason for worry. What to do with all this bounty?

Home-canning of tomatoes is an ancient practice all over the world. But, alas, cooking and stirring tomatoes for hours in a warm kitchen, with hot tomato juice spluttering on bare arms, is indeed no fun.

Let the Sunhood take over some of this tedious work. Only the customary preparations (washing, preliminary boiling) must be done in the kitchen, while the hour-long concentration of the purees and pastes is taken over by the sun. When the pureed tomatoes are poured into shallow pans, a natural evaporation of unwanted liquid will take place. No scorching can occur, and only infrequent stirring is required.

Paste, Catsup, Chili Sauce, etc.

Cut washed, unpeeled tomatoes into big chunks and boil until soft. Let them stand until the more solid parts sink and water rises to the top. (Remove this liquid, but don't discard it, for it can be used for soups or gravies. If this liquid is well-cooled it is a refreshing and healthful drink.)

Strain the solids or run them through a food-mill, and prepare them according to your own favorite recipe or use the following recipe:

Catsup:
>4 pounds tomatoes, unpeeled
>3 medium onions
>1 sweet pepper
>1 clove garlic
>Chop finely, or put through meat-grinder.

Tie in a cheesecloth the following spices:
>½ hot pepper
>1 bay leaf
>1 teaspoon mustard seeds
>1 teaspoon pepper corns
>1 cinnamon stick, 1 inch long
>Immerse in the tomato mixture.

Boil slowly, until vegetables are soft. Remove cheesecloth bag, and puree the mixture.

Add to the hot puree:
>¼ cup white sugar
>¼ cup brown sugar
>½ cup vinegar
>1 tablespoon salt
>Heat until sugar melts.

Pour the tomato puree in shallow black enameled pans and let the mild heat of the sun concentrate it to your desired thickness. When ready, empty the product into well-closing containers and sterilize. Invert the jars for added sterilization of lids.

Those tomatoes which are not used for juice, puree, etc., can be efficiently and safely dried under the Sunhood; they can be dried in slices, made into powder, or converted into Tomato Leather.

Tomato Slices:

Select deep red tomatoes with meaty flesh. Immerse them into boiling water for a few seconds so as to loosen the skins before peeling. Cut in ¼ inch slices. (Steaming the sliced tomatoes gives a product of somewhat better cooking quality.) Dry on plastic sheets, but remove the plastic as soon as possible, or use screen racks, made of a metal not attacked by acid. The screen should be treated with white neutral mineral oil. Dry until brittle.

Tomato Powder:

For powder, the tomatoes are merely sliced, de-seeded and dehydrated bone-dry, as described above. A few whirls in a blender will powderize the product. Keep the powder in well closed air-tight jars. Use for flavoring or for making tomato sauce.

Tomato Leather:

The tomatoes are treated the same way as for paste. Pour the puree in ¼ inch thickness on wet oiled paper and dry it until it forms a skin. Then peel the leather from the paper—while it is still warm—or it will not come off easily.

Plum tomatoes are cut in half for drying. Green tomatoes can be brought to a more speedy ripening under the Sunhood and then can be used with any of the methods given above.

INSTANT

SUN

RICE

One of the handiest staples to have in a kitchen is instant rice. After several trials and errors, I finally found the following way of making instant rice with sunshine.

Drop one cup of rice in plenty of boiling water, and boil it over high flame until the rice is done. The best test for this is to bite a grain in two. If there is no hard center, the rice is ready. Pour the rice into a coarse sieve, draining the water into a bowl (save the rice water, for it makes a good thickening for stews, curries, etc.). Wash the cooked rice with cold water until each grain is separated. Drain and dry on a clean towel.

Spread *thinly* on a baking pan covered with wax paper, and place it under the Sunhood. Shuffle the rice at regular intervals. Leave until bone dry.

This *instant sun rice* is in every respect compatible to any commercially available instant rice (only less expensive).

BEANS, LENTILS AND PEAS

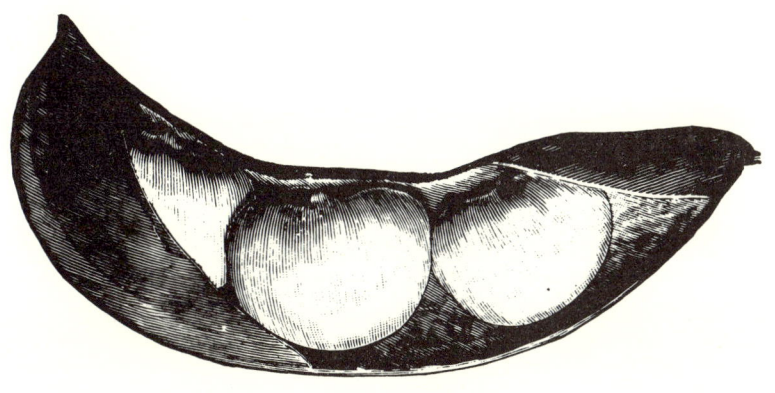

The dry, edible seeds of beans, lentils, peas, soybeans and other similar plants are commonly known as *pulses*.

It would appear that when these plants are so ripe that the seeds begin to drop, that they are then sufficiently (naturally) dehydrated by the sun for winter storage. But that is *not* the case. In order to prevent mold from developing on the shelled and sorted beans or peas, they need *further drying* under the Sunhood. This extra drying of the pulses also eliminates the need for fumigation, a practice used by commercial dehydration companies for killing weevils.

Harvest the beans on a dry sunny day, spread evenly under the Sunhood and dry until rock hard, shell and sort. Pasteurize in their containers before storage.

Lentils or Split Peas Soup Mix:
  1 cup lentils, or split peas
  2½ cups water
  1 medium onion, chopped
  ½ teaspoon thyme
  1 teaspoon salt
  crushed black peppercorns

 Drop the lentils or split peas slowly into rapidly boiling water, and boil for about 10 minutes. Reduce heat and add chopped onions and thyme. Simmer slowly for about 30-40 minutes, or until the vegetables are sufficiently soft to pass through strainer. Bring puree to a boil and add pepper and salt. For sun-drying—spread the hot puree immediately in ½ inch layer on shallow black pan(s). Use a fork when stirring. After mixture is completely dried, grind or crush, the solids to powder.

 Store in well closed plastic bags. Pasteurize the bags under the Sunhood for 30 minutes in bright sunshine.

Reconstitution:

 Soak the soup powder in water or vegetable broth (2½ to 3 cups.) Bring to boil and before serving add ½ cup consomme and ⅓ cup light cream and one teaspoon butter.

 Adjust seasoning. If the soup is too thick, add more consomme or milk.

Instant Boston Baked Beans:
Soak 3 cups dry kidney beans overnight and rinse with clean water. Blanch for three minutes in boiling water and place in a bean pot. Cover with warm water (about 1 quart) to which is added:

    2 teaspoons salt
    1 teaspoon dry mustard
    4 tablespoons dark molasses
    2 tablespoons vinegar

Bake slowly for 5-6 hours. Add water sparingly during baking so at the end only a small amount of liquid remains. Allow the beans to cool for 30 minutes, then load in two enameled baking pans at a rate of 1½ lbs. per square foot. Dry under the Sunhood until powder-dry.

Store in well closed plastic bag. Pasteurize the bags under the Sunhood for 30 minutes in bright sunshine.

Variations:
Split, green and yellow peas, as well as lentils, can also be converted into instant solar soups.

Roasted Soybeans:
Roasted soybeans are somewhat like roasted peanuts in flavor. They can be used in candy or eaten salted.

Wash and soak 1 cup dried soybeans in 3 cups of water for 12 hours. Change the water a few times. Drain.

Cover the soybeans with fresh water and simmer slowly for about 2 hours, adding water when needed. The beans are ready for roasting when they are soft, but not mushy. Drain thoroughly until dry. Blot off any remaining moisture.

Spread the soybeans on a lightly oiled baking sheet and roast in a moderate oven for one hour or until the beans are well browned. Sprinkle with salt, while they are still warm. Or, if desired, leave unsalted.

When the soybeans come out of the oven, they will be soft inside, but crunchy outside. After cooling, the soybeans become crunchy throughout.

Store in a cool place in tightly closed jars.

Other beans and dry peas can be used instead of soybeans. But soybeans are more nutritious, for they contain three times more protein than ordinary beans.

Nuts and seeds are roasted in similar ways in a dry pan in a moderate oven until they acquire desired brownness. Nuts and seeds no *not* require precooking, but beans or peas do.

HOW TO ENRICH FRUITS AND VEGETABLES

As the demand for dehydrated food increases, there often is also a demand for increased flavor.

With this in mind, I have for many years tried to find a method whereby anyone can transform simple products into culinary surprises.

With a little effort, it is possible to:
1. Enhance the flavor.
2. Augment the nutrient value.
3. Change the color of the product with natural ingredients. Briefly: *to produce unknown and unusual flavor combinations of fruits and vegetables.*

We can call this new approach of food drying the step-by-step method, because the product is treated in stages, as will be explained and illustrated.

BASIC

ENRICHMENT

METHODS

Step by step method of enriching dehydrated foods and improving their flavor.

Outdoors, Step 1:

The prepared fresh product is placed under the Sunhood, and dried to ½ or ¾ of its original volume.

During this stage, the heat of the sun drives the original water out of the fruits and the vegetables, leaving their cells partly empty.

Indoors, Step 2:
During Step 2, the partly dehydrated product is immersed into a fortifying liquid, or essence, consisting of fruit or vegetable juices in optional combinations. Spices and vitamins can be added to the essence, but no artificial flavors or colors.

The product should stay in this liquid for several hours, or overnight, or until it has plumped up to its original size and the cells have become loaded with the fortifying agents.

Drain if any liquid is left.

Outdoors, Step 3:
Place the drained and enriched product once again under the Sunhood for complete drying.

Vegetables should become bone dry, fruits until they feel leathery.

Store in sterilized and well closed containers.

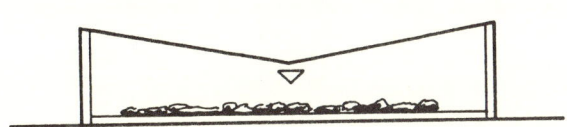

ORANGE APPLES

A step-by-step method of converting dried apples into a super-snack.

One single apple tree is able to carry a tremendous number of apples, which all ripen at the same time. What can one do with all this affluence? Obviously, it is impossible to eat them all. And adequate storage space is rarely available. Besides, some apples do not keep too well in storage.

One good solution to the problem is to dry the surplus. Apples are generally not sundried, because most varieties ripen in the fall, when there is little sunshine and it often rains. Apples are, therefore, usually machine dehydrated or dried in kitchen ovens.

However, under the Sunhood it is easy to dry apple slices even late in the autumn. In November, it may take a few days to get them dry, but in July, I have often dried apples in a few hours.

While I was trying to improve on flavor and texture of dried foods, one day I came across a new method by which it became possible to transform the otherwise rather bland apple slices into a delectable snack.

Step-by-step preparation of orange apples:
Select ripe apples of good quality: wash, peel, core, and slice them. In order to prevent the slices from darkening during the preparation, hold them in a weak salt solution(two tablespoons salt to one gallon water). Drain when ready to place them in the Sunhood.

1. Place the slices in a single layer on a rack under the Sunhood. Leave them until they become half dry (i.e. have lost half of their water content). If you have a scale, weigh them, or else judge the dryness from their touch.

2. Remove the partly dehydrated apple slices from the rack. Place them in a shallow dish; cover them completely with orange juice. Keep the dish in the kitchen and soak the apple slices for 24 hours, or until they have regained their original form and size.

3. After draining the slices, place them *once again* under the Sunhood, on the rack. This time, permit them to become completely dry. They will feel leathery and a bit sticky.

I admit that this method is a bit more cumbersome and takes a little more time than the ordinary, but it is well worth the effort and children simply adore the orange apples. Not only is it a healthy snack but it is also very inexpensive.

Other flavors:

Instead of orange juice, you can soak the apple slices in the juices of any citrus fruit: tangerine, limes, etc., either in pure juice, or in any combination with or without grated rind.

The juices of strawberries, raspberries, pineapples, etc., can thus enhance and change the flavor and color of otherwise rather bland tasting dried slices of apples or pears. You can easily create a multitude of interesting and unique variations.

Another method is to boil the peels, but not the cores or seeds, from apple trimmings in a small amount of water, until they fall apart. Strain the liquid through a cheesecloth and soak the half-dried apple slices in this apple essence. In this way the apples become fortified with considerable amounts of their own vitamins and minerals.

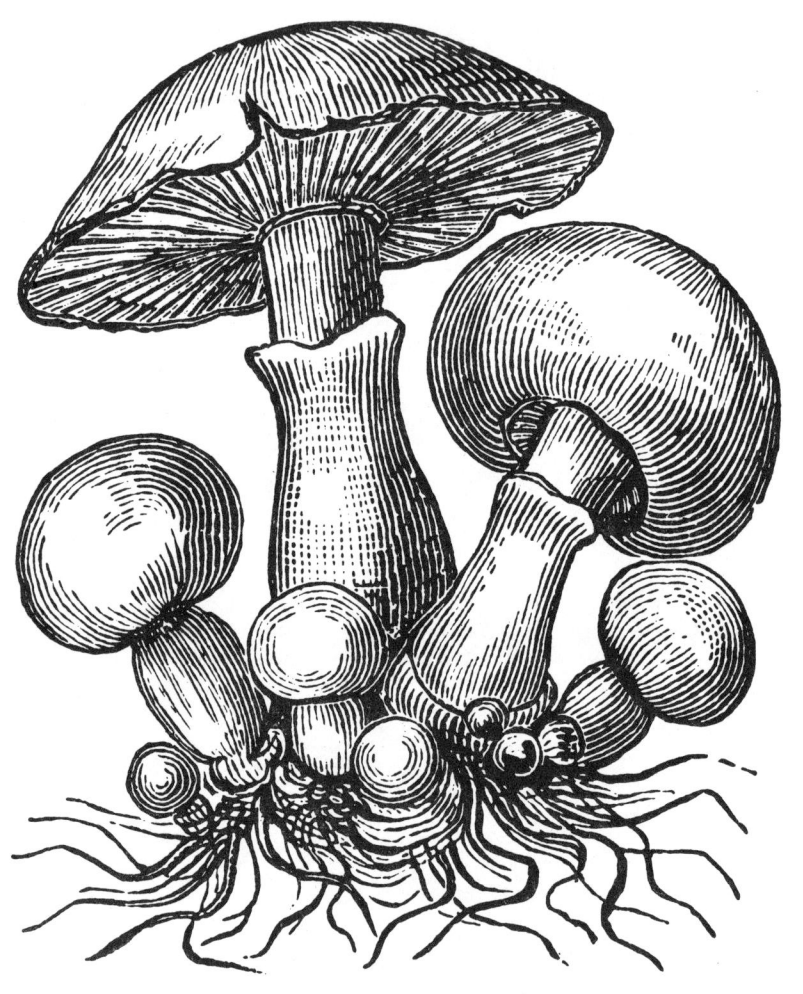

MUSHROOMS

Enriched by the step-by-step drying method.

If by chance you suddenly have more mushrooms on hand than you can serve your family and friends, then it is a good idea to dry this surplus by the step-by-step method.

Wild mushrooms (WARNING: Do not gather wild mushrooms, unless you are a mycologist, as some are deadly poisonous) have their own pungent flavor so they don't need any particular treatment before dehydration, but the forced champions lose even what little taste they ever possessed during the drying process. Even freeze-dried mushrooms have a peculiarly bland taste.

In order to obtain dried mushrooms with a great flavor, we suggest the following:

1. Wipe the mushrooms with a clean, damp towel, slice the unblemished mushroom caps and place them under the Sunhood (mushrooms don't require blanching).

2. Meanwhile, chop or grind the stems and blemished mushrooms, and cover with water and simmer slowly, covered, for 45 minutes or more. Spices and herbs like tarragon, basil, rosemary, etc. can be added to the water, or even onions, if desired.

3. When a sufficiently strong flavor is obtained, strain the liquid and let it cool. What you now have is a mushroom essence.

When the mushroom caps (which must be turned once or twice) have dried to half or two-thirds of their original volume, place them in a shallow dish and cover with the cold essence. Leave the caps immersed until they have regained their original size. Drain and as the final stage of the operation, spread the enriched mushrooms on the screen and dry under the Sunhood until brittle. Don't throw any remaining mushroom essence away, but use it to give additional flavor to a soup or sauce.

When reconstituted, the mushrooms can be used in any regular recipe.

The step-by-step method may appear slow and cumbersome, but it is well worthwhile because of the improved product. The point is that with the step-by-step method, you can "enrich" not only mushrooms, but you can load all kinds of vegetables or fruits with such minerals, vitamins, and flavors, which they previously did not possess. You can change the product and almost create a new variation of basic materials. The combinations and variations are almost limitless. Another great advantage is that no chemicals or artificial coloring or flavors are added, only nature's own ingredients.

Dried Mushrooms in Cream:

⅓ pound dried mushrooms
2 tablespoons butter
2 tablespoons finely chopped shallots
1 tablespoon flour
1 teaspoon lemon juice
½ teacup heavy cream
1-2 tablespoons sherry
pepper and salt to taste

Soak the mushrooms in warm (not boiling) water for 1 hour. Drain well and pat dry. Chop finely. Heat 2 tablespoons butter in skillet and add the finely chopped shallots. Cook briefly while stirring. Add the mushrooms, salt, pepper, and lemon juice, sprinkle with flour and cook covered over low flame for about 15 minutes. Gradually add the cream and cook for another 5 minutes, stirring constantly. Add the sherry and swirl in some more butter when ready for serving.

Creamed mushrooms can be served on triangles of toast, fried in butter. They can be used as filling in an omelet, or served in pastry shells.

Dill

SUN DRIED HERBS

The sunny corner of my garden, where my herbs are thriving, is one of my favorite spots. Here I grow my dill, parsley, rosemary, tarragon, thyme, basil, and many more herbs, which are used both fresh and dried in my kitchen.

In order to preserve their fragile and marvelous aroma, herbs should be picked early in the morning after the dew has disappeared. Wash gently and blot dry. Herbs do not need blanching. Spread evenly on the drying screen and place under the Sunhood. While drying, herbs should not be exposed to direct sunrays, therefore place Sunhood in open shade (see page 34).

Herbs are ready when bone dry and crumble easily. Crumble or powder and store in airtight containers

Dill:
Should be sown several times, because it is only the young green plants that are dried to "dillweed." The heads of the older plants are dried later when the dill seeds are ripe for use.

Rosemary and Tarragon:
The tips can be snipped off frequently. It seems that the plants become bushier and better the more often one snips off the tops.

Basil:
Should be picked before the flowers develop. I cut sprigs from the plants, which go on growing. The same plant can be harvested several times until the first frost kills the basil overnight.

Parsley:
A very valuable plant and should be used more widely than presently is the custom, it contains very beneficial properties—ounce for ounce parsley contains about four times more iron than spinach. It is richer in vitamin A than carrots, and leads lemon juice in vitamin C.

Thousands of tons of parsley are grown each year in this country, but only a fraction is ever eaten, for about 90% is used merely for decoration, whereafter this garnish is dumped into the garbage can.

Parsley should be used more freely in soups, stews, sauces, salads, meatloaves, etc. I always dry a lot of parsley to have on hand when winter comes.

Basil

MAKE YOUR OWN HERB MIXTURES

Poultry Seasoning:
    Consists generally of equal parts of: dried sage, marjoram, thyme, and savory to which a dash of spices like pepper, nutmeg, and allspice can be added. The dry mixed herbs are crushed or powdered and stored in small tightly-closed jars.

Herbs de Provence:
    Is a mixture of herbs for gourmets, but it is quite easy for anybody to become a French chef by making the mixture yourself.
    Mix well the following finely powdered dried herbs:

Basil
Thyme   1 tablespoon of each
Oregano
Marjoram

Rosemary
Savory   ½ teaspoon of each

2 well-crushed bay leaves

Store in small labeled and tightly-closed glass jars.

Herb Flavorings:

During drying, vegetables lose part of their fresh flavor. The careful addition of spices and herbs like parsley, dill, and basil improves the taste. Dried yellow peas go well with a little thyme. Split or whole green peas with mint and onion. Dried carrots should be cooked with a little sugar and butter in the water and flavored with dill or parsley. Tomatoes go well with onion and basil, and red beets with horseradish and carroway seeds.

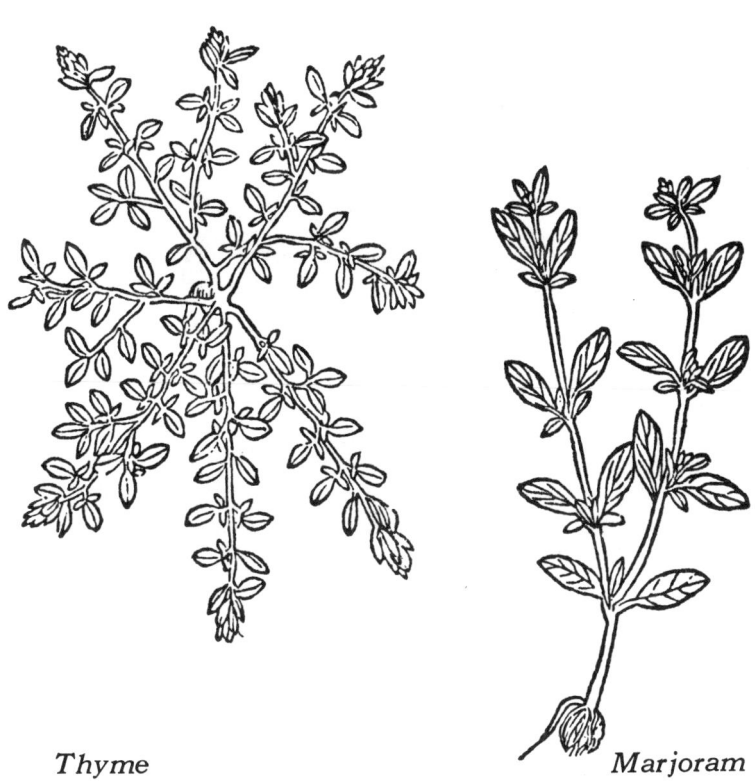

Thyme   Marjoram

HERB TEAS

In Europe during World War II, we had to use all sorts of "ersatz" products among those were, *herb teas* of which I tasted a great variety. Some were quite pleasant; others were truly awful. However, tea mixtures which I found very agreeable were the following.

Strawberry tea:
Mix: equal parts of dried leaves of wild strawberry, raspberry, and blackberry. Dried blueberry leaves are sometimes added.

Linden tea:
Belongs to one of my favorite tisanes, served with a slice of a lemon and a spoon of honey. Linden tea is not only tasty, but it is recommended for those who have weak stomachs or suffer from colds. One drawback is that as linden trees grow so tall, it is often quite a problem to pick the flowers. Nevertheless, it is worth the effort to obtain them, for home dried linden tea is so much better than any store-bought variety.

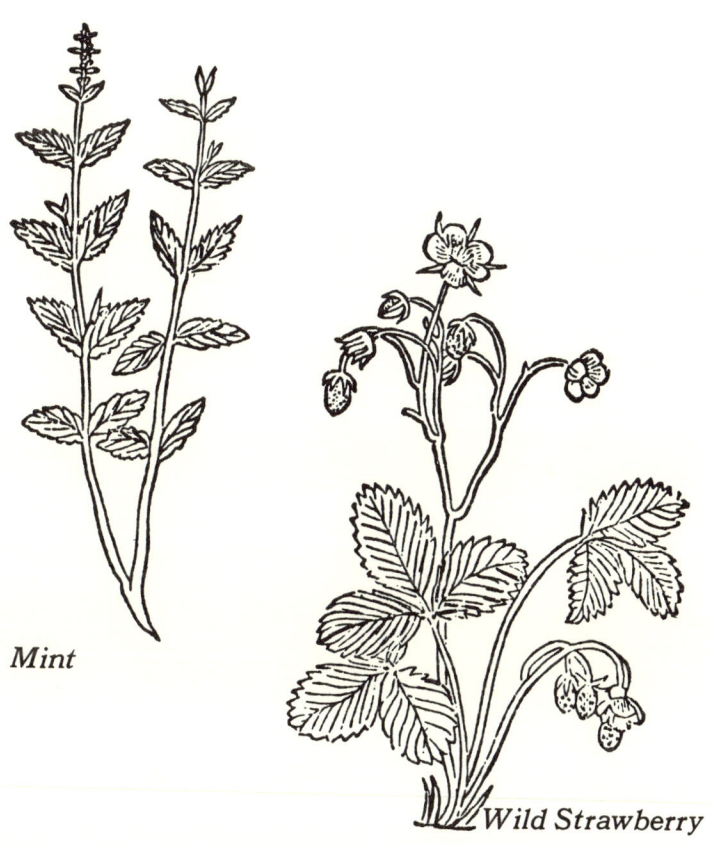

Mint

Wild Strawberry

Mint tea:

Around the well in our little garden grows mint in great abundance, which we harvest several times during summer and autumn, and dry for winter use. I pick the whole tops and the leaves, stripped from the coarser stems. The dried mint leaves are stored in well-closed plastic bags. For many years, mint tea has been our after dinner beverage. It is pleasant, healthy, and inexpensive!

CHAMOMILE FOR BEAUTY AND HEALTH

For centuries this little plant, which looks like a diminutive marguerite, has been known and treasured for its healing and cosmetic properties.

It grows wild all over, but can also be grown in any garden. It thrives in hot and dry places where nothing else will prosper.

I always pick the flower heads about noon on a sunny day, when they are in full bloom. In Hungary, where they grow in abundance on the prairie, I learned how to pick them with a comb. (Place a comb underneath the blossom and lift.) This method is much quicker than pulling off each little flower head by hand.

After picking, spread them out thinly under the Sunhood and dry in open shade. The dried product must be kept in well closed containers or double plastic bags.

Chamomile tea has a soothing effect on the nerves and is an excellent aid to digestion after a heavy meal. For a single cup of tea, use a heaping teaspoon of the dried chamomile flowers in a cup of boiling water. Let steep for 3 to 4 minutes before straining. Sweeten with honey and include a few drops of lemon juice.

As a cosmetic, infusions of chamomile flowers have been used for centuries by women for beautifying the skin, and to preserve and give a golden tint to blond hair.

For a hair rinse, place 4 tablespoons of the dried flowers in a pint of water. Boil for 30 minutes. Strain through a towel and let cool. Use as a rinse after the hair is shampooed. An old trick is to wet and rinse the hair several times with this mixture, and dry the hair in full sunshine after each rinse.

But flowers distilled,
Though they with winter meet,
Lose but their show,
Their substance still lives sweet.

     Shakespeare

STORING THE FRAGRANCE OF SUMMER FOR THE WINTER

During those days when the Sunhood is not needed for garden products, let it serve the enjoyable purpose of drying the fragrant petals of roses, lavender, honeysuckle, carnations, and whatever sweet smelling plants you have in your flower and herb garden. After drying, the petals are stored for potpourris or in sachets.

The petals should be shuffled daily until dried. Usually, only flower *petals* are used for this purpose, but I always add a few small rosebuds as an added accent.

It is important whenever one dries flowers or herbs, *not* to keep the Sunhood in direct sunshine, but to place it in open shade.

When perfectly dried, store the petals in decorative potpourri containers, or in well-closing jars. I found the best way of making a potpourri is to spread the dried petals thinly in the container and to sprinkle salt between each layer, together with spices, like powdered cinnamon and whole cloves. Small amounts of dried rosemary and mint leaves give an added accent to the mixture. As different flowers are ready to be picked, new layers can be added to the previous mixture.

Keep the container well closed, but stir the potpourris daily until all fragrances have blended.

During the dark winter months whenever you open the lid of the potpourri jar your room will be filled with the fragrance and sweet memories of the summer past.

Sachets:

Another way of putting fragrant flower petals to use is to stuff them into small sachet bags made of sheer fabrics. Close the sachets with pretty ribbons. When placed in your drawers, or in the linen closet, these little sachets will continue to give off their scent for years to come. In addition, they make elegant gifts for all occasions.

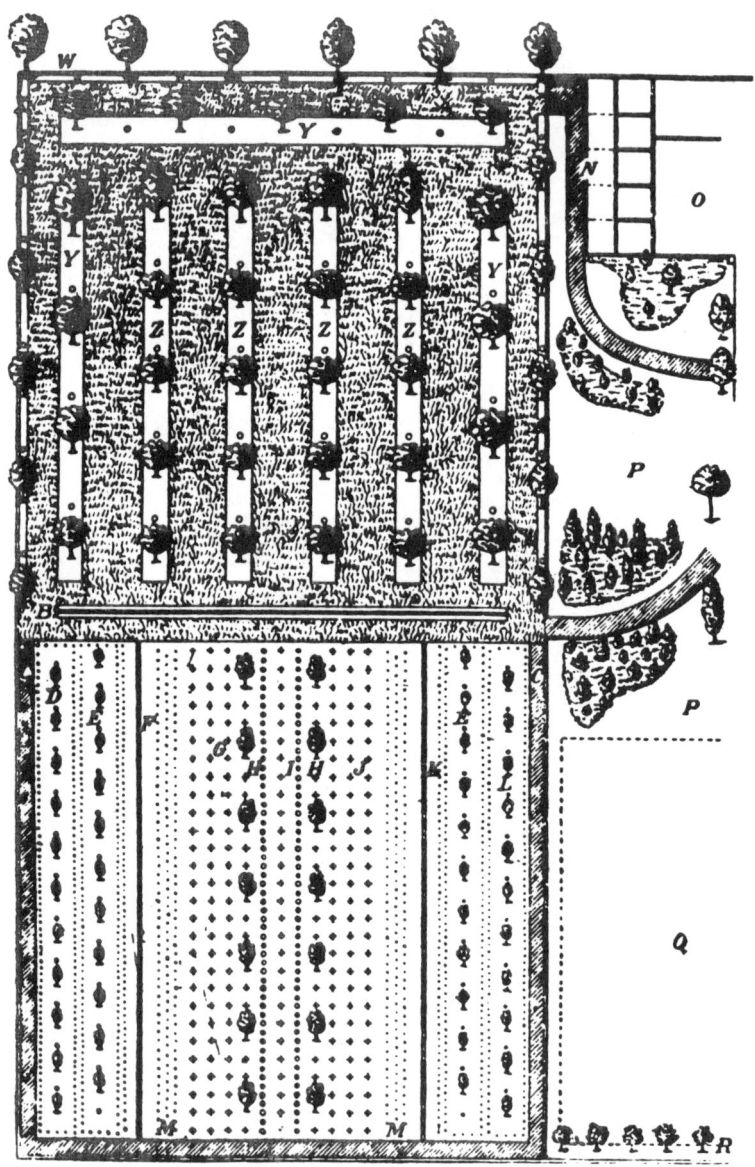

USE THE SUNHOOD AS A MINI-GREENHOUSE

Place the smallest seedlings in peat-cubes or eggshells in the center of the Sunhood. Put the higher pots or seed-flats along the sides. To keep the humidity high, the drip-channel can be removed. Ventilate, when needed, by lifting up one side of the mini-greenhouse.

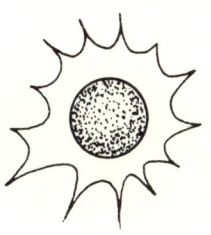

FINAL WORDS

The conclusions which can be derived from over a quarter of a century's experience with solar drying of fruits and vegetables, clearly indicate that the preservation of foods with controlled solar energy is not only feasible, but also *profitable,* mainly due to curtailed losses and improved quality of the products.

It is profitable, not only for the small garden and orchard owner, but even more so for those commercial dehydration operators who still practice large scale open air drying, where a large portion of the valuable crops are constantly destroyed by scavengers and unfavorable weather conditions.

The research and development of industrial solar driers deserves, indeed, much more attention and wider support from the big fruit dehydration corporations, than hitherto has been the case.

DO NOT USE CHEMICAL FOOD ADDITIVES

Many people object to the use of chemicals in food, and for that reason I nowhere recommend sulfuring, lye-dipping, or fumigation, nor washing in a solution of hydrochloric acid, etc. Personally, I consider the use of chemicals in this connection too dangerous for laymen to handle.

I have found that, even if the apples and apricots that I dry without sulfuring, are not as white or beautiful as the store bought ones; they make up for it, for they have a better taste.

DO NOT SUN DRY MEAT OR FISH

Nowhere do I mention the sun drying of meat, shrimp or fish. This is an art in its own right, and cannot be learned from a book, but only from somebody with long practical experience in the field. WARNING: Any mistake in the handling of these perishables could lead to food poisoning.

COMPARISON OF SOME FOOD DRYING PROCESSES

1. *Sundrying in the open air,* as it still is practiced around the world, consists of placing the fruits or vegetables on racks or trays, or directly on the ground. Here the product is exposed to dust and dew, birds and insects of all kinds. Thus, even if the fuel is free, it is a very uneconomical process. According to statistics, in many instances, as much as twenty-five percent or even more of the product can get destroyed during one season. A long period of rain, during the drying time, can easily wipe out the hard-earned profits of the grower.

2. *Machine Dehydration* is, in comparison, much more hygenic, and it is also much faster than the open air drying. However, large investments are required for equipment, and fuel expenses are high. Machine dehydration belongs, therefore, within the realms of the big industrial food companies, and is out of reach for the small farmer and home gardener.

3. *Food drying with the Sunhood* represents a combination of the two above described methods:

    a. As in the open air drying, *the fuel is free.*

    b. The *losses,* due to spoilage and unfavorable weather conditions, *are very much curtailed.*

    c. The product is *equally as sanitary* as its machine dehydrated counterpart.

    d. The drying time is considerably shorter than in open air drying.

RESULTS OF SOME WINTER TESTS

While November and December would not be considered an ideal time to sun-dry produce in many parts of the United States, I did conduct some drying tests with the Sunhood during these months in New Jersey, with the following results:

Food	Time
Broccoli	3 days
Carrots	18 hrs.
Celery tops	3 hrs.
Coconut flakes	4 hrs.
Cranberry halves	18 hrs.
Dill	5 hrs.
Grapes	8-14 days
Onions	24 hrs.
Parsley	5 hrs.
Peaches	5 days
Peppers, hot, long	12 days
Pimento	30 hrs.
Plums	48 hrs.
Seaweed	9 hrs.
Spinach	2½ hrs.
Squash	18 hrs.

RECORD KEEPING CHART

Location _____ Date _____

Type of Food	Quantity	Time		Weather
		In	Out	

Details of Food Preparation:

Remarks on Results:

RECORD KEEPING CHART

Location _____ Date _____

Type of Food	Quantity	Time In	Time Out	Weather

Details of Food Preparation:

Remarks on Results:

RECORD KEEPING CHART

Location _____ Date _____

Type of Food	Quantity	Time In	Time Out	Weather

Details of Food Preparation:

Remarks on Results:

RECORD KEEPING CHART

Location _____ Date _____

Type of Food	Quantity	Time In	Time Out	Weather

Details of Food Preparation:

Remarks on Results:

APPENDIX

The Sunhood described in this Solar Food Drying handbook is designed so that a single person can handle it easily.

With the present size, three to four pounds of vegetables or fruits can be dried with one loading. In case you have a large garden or an orchard with an abundance of fruits—it is advisable to build *several* Sunhoods.

The Sunhood is not limited to a single size. It can be made to fit any product and can readily be adapted to large scale solar drying.

Write to us if you have a special problem, and we will be glad to discuss your questions.

A Sunhood Kit is available and can be ordered from Solar-Electric Laboratories, Box 385, Kingston, N.J. 08528, or for information write the publisher.

RECOMMENDED READING

Regardless of whether one dries food in tunnel or drum-dehydrators, or applies vacuum or freeze drying methods, or if one uses solar driers, all the preliminary precautions and preparations are of the same nature. The drying of foods is a serious science and there exists an extensive literature on the subject. Among the books I have studied most carefully and used as guidelines during my research of 25 years, are the following:

Raisinland U.S.A., California Raisin Advisory Board, Fresno, California.

Preparation of Cocoa, Caribbean Commission, Trinidad, Publication Exchange Service, No. 60.

Preservation of Fruits and Vegetables, by E. M. Chase, US Department of Agriculture, Bulletin 619, 1942.

The Dehydration of Vegetables, by W. V. Cruess and E. M. Mrak, University of California, College of Agriculture, 1941.

Drying Fruits and Vegetables, by Inez Eckblad, Colorado State College, Bulletin D-47.

Unsere Heilpflanzen, by Eugene Fischer, Albert Mueller, Verlag Zuerich, Switzerland, 1941 (in German).

Valsignade Vaxter, by Nils Hewe, Natur och Kultur, Sweden, 1947 (in Swedish).

Soybeans for Health, Longevity, and Economy, by P. S. Chen and H. D. Chen, The Chemical Elements, South Lancaster, Mass., 1956.

The Possibilities of Using Solar Energy for Drying of Fruits and Vegetables, by A. A. Ismailova, USSR. Paper presented at the Solar Symposium in Phoenix, Arizona, 1956.

Dehydration of Meat, by H. R. Kraybill, 1935, Ind. Eng. Chem.

Drying and Dehydration of Foods, by H. W. Loesecke, Reinhold Publishing Corporation, 1943.

Common and Uncommon Uses of Herbs, by R. Lucas, Parker Publishing Co., 1969.

Methods and Equipment for the Sundrying of Fruits, by Mrak and Long, University of California, Bulletin 350.

Sun-Drying Fruits, by Mrak, and H. J. Phaff, University of California, Bulletin 392.

Fruit Dehydration, Perry, Mrak, Phaff, et. al., California Agricultural Experimental Station, Bulletin 698, 1946.

Sulfur House Operation, Phaff and Mrak, California Agriculture Experimental Station, Bulletin 382, 1948.

Sun Dry Your Fruits and Vegetables, Helen Straw, US Department of Agriculture, 1958. For Home Economists around the World.

A Brief Summary of Activities of the U.S. Department of Agriculture in Dehydration. Month. Bulletin. Dept. Calif., 9, 1920 by P. F. Nichols.

Putting Food By, by Ruth Herzberg and Beatrice Vaughan, The Stephen Greene Press, 1975.

The Complete Sprouting Cooking Book, by Karen Cross Whyte.

How to Sun Dry Fruits and Vegetables, AMBIT Enterprises, 1975.

Build a Solar Cabinet Dryer for Year-Round Profits, by Steve Smyser, Organic Gardening, December, 1974.

How to Dry Food at Home, EVP Enterprises, 1975.

Add Water and Wait, by Susan Margolies, Wall Street Journal, March 5, 1975.

Starting Seeds Outdoor in Cold Frames, by Karman McReynolds, Organic Gardening, February, 1976.

Magazines such as *Mother Earth News, Organic Gardening, Country Living* and *Harrowsmith* (Canada) often run articles on drying produce.